THE MANHATTAN PROJECT

THE MANHATTAN PROJECT

KEN HUNT

Brave & Brilliant Series
ISSN 2371-7238 (Print) ISSN 2371-7246 (Online)

University of Calgary Press
2500 University Drive NW
Calgary, Alberta
Canada T2N 1N4
press.ucalgary.ca

LIBRARY AND ARCHIVES CANADA CATALOGUING IN PUBLICATION

Title: The Manhattan Project / Ken Hunt.
Names: Hunt, Ken, 1991- author.
Series: Brave & brilliant series ; no. 14.
Description: Series statement: Brave & brilliant series ; no. 14
Identifiers: Canadiana (print) 20200212478 | Canadiana (ebook) 20200212540 | ISBN 9781773850542 (softcover) | ISBN 9781773850559 (PDF) | ISBN 9781773850566 (EPUB) | ISBN 9781773850573 (Kindle)
Classification: LCC PS8615.U58 M36 2020 | DDC C811/.6—dc23

The University of Calgary Press acknowledges the support of the Government of Alberta through the Alberta Media Fund for our publications. We acknowledge the financial support of the Government of Canada. We acknowledge the financial support of the Canada Council for the Arts for our publishing program.

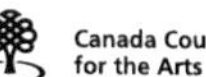

Printed and bound in Canada by Marquis
♻ This book is printed on Lynx Opaque paper

Editing by Helen Hajnoczky
Cover image: Colourbox 12789774 and 23995089
Cover design, page design, and typesetting by Melina Cusano

THE ATOMS WE CLEAVE

Agitated particles invade a waning forest.
Brittle pines, tainted by an errant strain

of iodine, rise like headstones from the
quarantined graves of their progenitors.

Wolves, all mange and sinew, prowl gutted
towns, stalk hollow barns stuffed with cattle

bones, sniff the pews of churches whose
sagging walls now bow to a new god.

Artists turn ruins into canvases. In deserted
power plants on the Jägala River, and in decrepit

Pripyat apartments, they stage games of musical
chairs with ghosts, hoping to catch a perfect shot.

Incandescent nymphs await lost authors, wayward
pilgrims that wander within range of their impish

hymns. In deep hives, tribes of caustic sprites curse
the poisons that birthed them. Songs of damnation

drone from their glowing throats, drawn from
the syntax of deep time and put to notes that shame

the sirens' calls. Ensnared artists retreat into
the sickly glow of trees imbued with virulent auras.

The nymphs' eyes shimmer with malignant
minerals, the dusts of dead suns. Their enriched

kisses transmute flesh into a photo of its shadow,
dissolve lithe bodies subatomically in acts

of transcendant decay. The sisterhood
of nymphs make pilgrimage from their buried

cells to the throbbing graves of "biorobots,"
Soviet conscripts who shovelled the shredded

entrails of graphite control rods from the caving
roof of Chernobyl. Years ago, the split reactor core

became a portal to the scabbed magma
of adolescent earth. The nymphs offer prayers

to the restless dead. Beyond thier lairs lie acres
of necrotic grass, where two-headed lambs graze

clumsily, each head tugging its shared neck
in a different direction, toward the bite the other

head can never see. Geneticists experiment,
implanting a spider's silk-producing genes

into lamb cells, in order to milk the newly-made
chimera for organic fibers with a tensile strength

greater than steel. Beyond the fields, and beyond
the meadows and the tree line, in the heart

of the Red Forest, a troupe of hairless faeries
guard a hidden glade. They hover, scouting

its perimeter, to protect the tree of death, whose
twisting branches sag with tumorous fruits, their

skins encrusted with heavy metals, their seeds
aglow and showing through. Grisly gnomes huddle

in burrows around the roots of this mammoth clot
of bark. When the moon's gaze falls behind the clouds,

the dead climb out of the burls of the tree, exposing
the hoaxes of all clocks. The sun, in protest, rises

from the wrong side of the sky, a burst of dying light
from the depleated heart of an angel delirious

with weariness, not cast down, but lost like Dante,
in the Red Forest, tormented by the depths of mortal

suffering. The nymphs laugh, while the angel, blinded
by the glare of humanity's false suns, begins to weep.

THE ARMS RACE

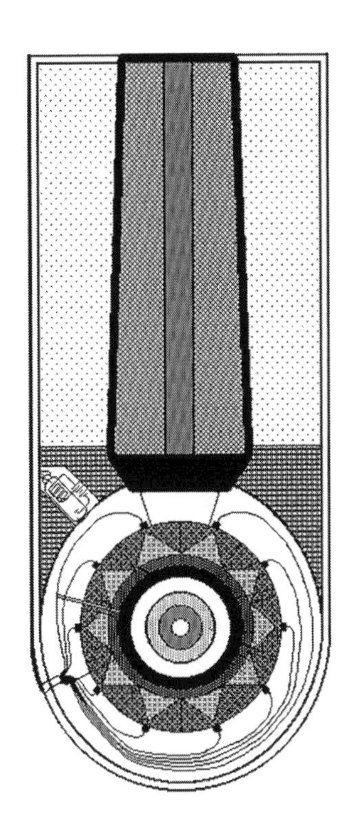

BELOW OKLO

Press your ear against this fossilized nautilus
to hear the hum of a natural reactor.

Below and before the colonial mines, before
the bombs, the fallout, and the shelters,

there was a buried decanter of light, a lair
of drakaina, a chasm where a granular fuzz

of uranium crystals tickled the feet of eyeless
naiads, their silver skin dipped in a balm

of stray ions. These nymphs bathed
in superheated cisterns of trapped water.

They fed on plumes of heat produced
by buried suns, whose pungent rays pickled

the tissues of the earth. They drank
the dew of Styx from crystal goblets.

Ancient reactor coolant pacified
the shrieks of stillborn stars, whose songs cut

through the earth with wild notes, each burst
of fission sizzling like a sunken lantern

plunging into the maw of an subsea trench.
The naiads' infernal sauna predates us;

this Pandora's box unlocked itself. Before
our earliest ancestors first tread through

southern savannahs, a restless trove
of nuclear fuel pulsed in this georeactor,

each Precambrian throb a spasm of ore,
a radionuclear twitch eager to spill forth.

Neodymium dissolves in this aching heat.
Ruthenium unravels in this raving deep,

decay particles caught in sandstone,
clay, and granite. Thermal neutrons

sunder the surrounding umber stone
of these hothouse catacombs.

Carcinogenic steam from hellish
bathhouses permeates troughs

of liquid heat, where even molecules boil,
where even nuclei evaporate.

In the naiads' company, a necromancer
charms the cavern's dead back to life.

Calcified skeletons crack open their
stratified tombs to dance

in the antechamber of Earth's
first critical mass.

Nature was never innocent, trapping hymns
within black crystals, testing her flesh

in water-woven trenches, breeding grounds
for her tectonic fauna: uraninite, pitchblende,

thorianite, pegmatite, betafite, lost volumes
from a mineralogical apocrypha.

The demise of our species
is written in these stones.

There was a revelation when the mines opened,
though the miracle was merely material.

Plunderers dove into the earth
for the spoils of energy. And human life

prepared itself for omnicide,
bathing in the waters of its doom.

RADIOACTIVITY

for Marie Skłodowska Curie

Relentless curiosity compelled you to plunge
your hands into elemental embodiments
of chaotic decay, to tinker with glinting flasks
of vicious species of dust.

If young Joan of Arc spoke with god, and was burned
for their exchanges, then which gods communed with you
and set your bones ablaze, left you delirious
from necrotic marrow?

For how many hours, O bright priestess
of Prometheus, did you bear bundles
of test tubes, slid into the pockets
of your lab coats, each glass vial alight

with flameless fire, leaving your clothing unsinged,
yet malignant? The fraying atoms you interrogated
co-wrote entries in your journals, embedded
marginal notes between molecules of ink.

Reading these pages of cursed prose
now requires protective gloves. Samples
kept in the drawers of your desk gave off
faint auras, your will-o-wisp companions

during winters spent purifying powdered ores.
As Red Cross director, you drove a mobile
x-ray cart across fields laden with corpses
and scorched iron, mending the bones

of wailing soldiers. You sterilized their wounds
with radon needles, but you couldn't save
the book of recipes you cooked your every meal
from, a book now kept in a lead-lined box.

Some time after the trauma
of your husband's death, did you recall
a former lover, named Kazimierz Żorawski,
a mathematician whose parents forbade

their son associate with a penniless girl?
Years after your death, a certain old professor
could be seen each day, seated before
the statue of you, erected

at Warsaw Polytechnic, where he lectured.
The arcane weight of artefacts
mangles the steady gaze of history,
overexposes fantasies of clarity

with scathing rays. What more is there,
other than chemistry, for writing to occupy?
What meagre certainties can we attach
our pens to? The measure

of a half-life is subject to estimates, imperfect
measurements, and unchecked variables.
No one can see what has been,
nor what is left to be.

THE WORLD SET FREE

for Leo Szilard & Otto Hahn

The cry of a multitude, screaming
in a lucid nightmare of light, echoes

backward, into 1933, but the voices fail
to out-sing the crinkle of dying rainfall

on a brittle London street, where Leo Szilard
walks (incensed at Lord Rutherford's

brazen dismissal of his work) and sees
the tree of death erupt before him,

from the seed of a buried dream.
Did he envision what repercussions

would arise from the plucking of its fruit?
Was it really only in that moment at Farm Hall

in 1945 when, as a comfortable prisoner
of Operation Epsilon, he listened with

his colleague Otto Hahn to the radio broadcast
that proclaimed the dropping of the bomb,

that his vision broadened, stretching past
the border of complicity? How could grim doubt

fail to haunt Hahn when, in 1916, he was stationed
in Fritz Haber's chemical warfare unit

as a researcher of cytotoxic poisons?
A year later, at the battle for Hill 70,

the faces of several soldiers bore
new varieties of shrapnel, such as

bubbles of chlorine gas that got trapped
behind their eyes. In 1938, when Hahn gave

his mother's diamond ring
to his Jewish pupil Lise Meitner, as a bauble

to bribe a German border guard and guarantee
the girl's escape, did that act not embody

the writhing of his mournful ghost, condemned
to spend eternity emptying his depthless pockets

searching for spectral coins in vain, gleaning
no balms from his desperate gestures of atonement?

Perhaps Szilard had some inkling,
in the fatal stillness of that morning, waiting

to cross the street, that the eyes of time were on him,
and no kind of judgment, even if it came,

would scrub his memory of that crossing
over ragged pavement slick with rain.

IDEAL ISOTOPES—PRELUDE

for Enrico Fermi

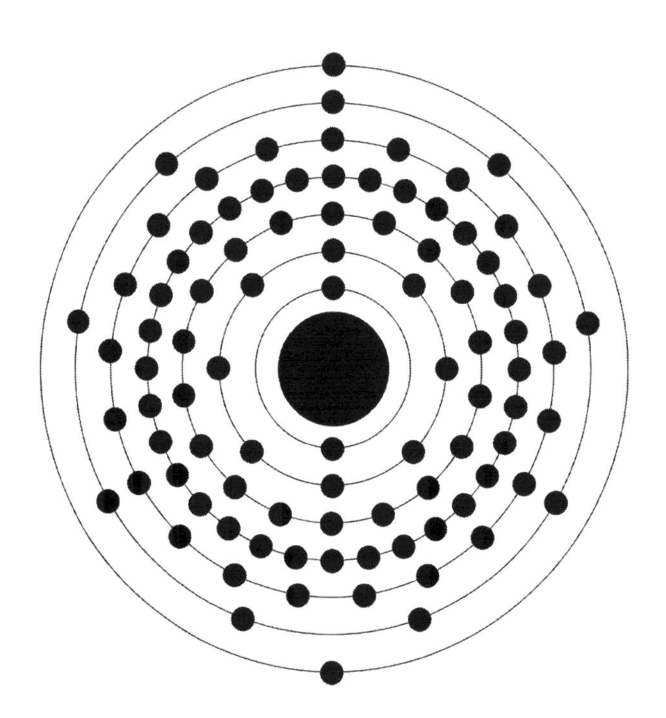

fig. 1: diagram of an atom of U-235

URANIUM 235

Uranium is an actinide metal that appears naturally in ores such as uraninite and torbernite. Approximately 99.3% of natural uranium atoms have nuclei made up of 92 protons and 146 neutrons, making their atomic weight 238. The remaining 0.7% of uranium found in the earth has only 143 neutrons to compliment its 92 protons, culminating in an atomic weight of 235. This second, rarer type of uranium atom is called an isotope (a fusion of the Greek root terms isos, meaning 'equal', and topos, meaning 'the same place'). These isotopes of uranium 235 are less stable than uranium atoms containing three more neutrons. Because of their instability, atoms of U-235 are prime candidates for atomic fission, or the shattering of heavy atomic nuclei which results in the release of high-energy neutrons, which in turn shatter the nuclei of nearby atoms, resulting in a chain reaction that produces immense amounts of energy. A series of concentric 'shells' of electrons surround the nucleus of U-235, attracted to the positive charge of its nucleus. Figure 1 illustrates the U-235 atom, with its massive nucleus and haze of orbiting electrons. Distilling U-235 from the more common U-238 is difficult, since the uranium atoms are each nearly identical in chemical behaviour.

We yank the deadly reins of the four horsemen when we seek to enrich the rage of metals dashed into black crystals, coagulations of chaos that disrupt fragile DNA.

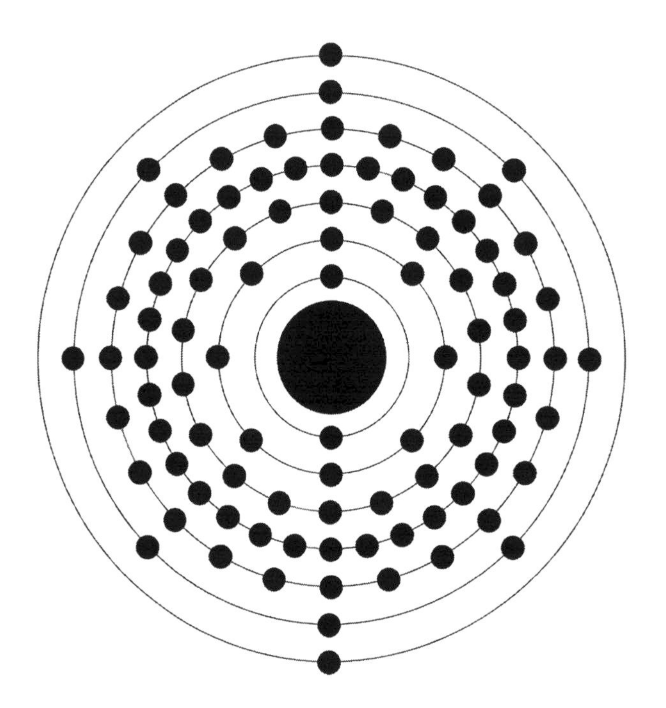

fig. 2: diagram of an atom of Pu-239

PLUTONIUM 239

Four isotopes of the actinide metal plutonium occur in extremely small quantities in uranium-rich ores, as byproducts of uranium's decay. The vast majority of plutonium on Earth is produced in laboratories. Bombarding uranium 238 with neutrons initiates a reaction which can produce the isotope plutonium 238. Further bombardment of Pu-238 with neutrons produces Pu-239, the most viable isotope for nuclear fission. Pu-239 contains 94 protons and 145 neutrons, making its atomic weight 239. Pu-239 is cheaper to produce in large quantities than U-235, which has led to the element's use in both nuclear weapons and nuclear power plants. The critical mass of Pu-239, or the minimum amount of the isotope needed to invoke a nuclear chain reaction, is the smallest of all nuclear fuels. An 11kg sphere of high-grade Pu-239 is enough to initiate such a reaction, although certain processes can reduce this amount by half. 'Supergrade' nuclear fuel contains 95% or more Pu-239, the remainder being Pu-240, and is used in situations where lower radioactivity is necessary, such as submarines, where crews operate in close proximity to nuclear weapons. Just as uranium is named after the planet Uranus, plutonium is named after the dwarf planet Pluto.

We tickle the tails of dragons to animate our forges with their flames. We prod the hearts of demons to decipher their rhythms, only to have our own tissues unwoven.

IDEAL ISOTOPES

Acheron, the river of woe, flows past
Charon's abode. Nearby, gold wolves prostrate
before their alpha Anubis. A raft

awaits Aeacus, who describes the fate
of his grandson Achilles to A Bang, who,
as Bull-Head master of a grand estate,

held a banquet for the god 'A', whose crew
of Mayan wraiths resent their extinction
at Spanish hands, each conquistador doomed

by Coatlicue, the snake-skirted matron
of the moon, to live the final moments
of their victims endlessly. Aminon,

gatekeeper of the Ossetians' trench,
wagers that Apialoovik can't outswim
Tlalok, and watches as their argument

erupts. Yonggung Sacha enters the din,
insulted by exclusion, while Aita
sighs, bearing witness with his unseen twins

Hades, Pluto, Orcus, and Dis Pater
from their tower above the foam of souls
and clash of bladed fins. Old Barastyr

scolds his servant Aminon for goading
the seabound foes, while the lion Aker
rears at Alpheus, who was washed ashore

by the nautical scuffle. From the blur
of tar-black waves, Andjety's hand swept forth
to grasp the gods bickering in the ichor.

Aqen, the mouth of time, rallies his throes
of allies at Andjety's rise: Arawn,
and Anguta the Mori, Atropos,

Angelos, Angra Mainyu, Ban Jian,
Bai Wuchang, Chen De and Cheng De, Ankou,
and Chitragupta. Cheonjiwang glares

into a mirrored wall of iron-blue
vivianite, at forces overdue
for war and summons, from the glass, a noose

to rival the rope of Asto Vidatu,
used to catch human spirits trying to flee.
Azrael, startled by this act, calls to

his kin Bao Zheng, and Barons Samedi,
La Croix, and Cimetiere, to wake their droves
of sleeping banshees and black dogs. Dark seas,

now greased with godly blood, disturb Cichol,
whose benthic groans startle Cao Qing, Bian Shen,
Chen Xun, and Cai Yulei. They loathe Clotho,

who quit the fates for his friend Jiang Ziwen,
to aid Jiang's bride Cihuateteo, along
with Chepi and Cheonha Daejanggun.

The latter two hide Ziwen's precious one
beneath where Cocytus's waters flow,
where she bids the Erinyes to summon,

with shrieking song, St. Patrick's primal foe
Crom Cruach, who resides in Bull Rock with
The Dark One, Donn, guarding the sacred souls

of faithful Gauls, who once fought with Cu Sith,
the giant wolf whose cries could split the moon.
Chief Judge Cui decrees martial law, and with

his roar brings forth the judge Dong Jie, with whom
King Dong Ji nearly shares a name. Culga
claw apart their crypts, screeching in the gloom

their neverending nocturnes, as the bulge
of the snake Degei's girth erupts, a drum
beat echoing in the volcanic gulch

where choirs of Di Inferi still hum.
Duamutef bellows atop his trove
of canopic jars, calling to Cui Cong,

who invokes Erebus and El Tío,
whose march disturbs Hel, Djall, and Diao Xiao,
locked in an orgiastic fit below

the knoll that Ghost King Duzi Ren calls home.
Dullahan, Eridanos, and Gao Ren,
out hunting near the knoll, awake Fu Po,

whose thrumming yawn inspires the Guédé
to launch into a polyrhythmic spree
with blazing drums. Erio and Guaiwang

share a frenetic waltz before Hapi,
who whispers to Februss of disarray
in the realms of death, a rampant disease

of war heretofore unseen. Sly Freyja
eavesdrops. She warns Daebyeol of a con,
bids Ereshkigal and Danmul Sacha

spy on King Fan Zhongyan, inspect this pawn
in the unfurling chaos. The King's own
spy, Han Yi, hears this and alerts Guo Yuan,

who arranges with Gangnim Doryeong,
Giltinė, Gorgya, and Hine-nui-te-pō
to raise a mercenary force. Along

King Han Qinhu's border, a wondering soul,
the damned mystic, Fan Wujiu, chants a tune.
Spectral troops flow from Khagya-Yerdi's knolls

to pierce the distant dark. High in this gloom
the Grim Reaper and He Wuchang fast drain
an obsidian carafe of baiju

at a diamond table in Mot's chalet.
The vapours from the liquor waft and swirl,
gossamer threads of reaped abyssal grains,

fermented near the Horned God's fields, where
pungent harvests fall to Menoetes,
Lemures, and Lamia. Their crops bear

fungi, lichen, mushrooms, and mosses, blessed
by rot, inspected by Hunhau. Kou Zhun
handles fermenting mashes he has pressed,

while Molyz-Yerdi, Liu Cha, and Jia Yuan
taste each new batch from trinitite snifters
before approving shipment to the vaults

of Iku, Mannanan, Libitina,
Liu Bao, Izanami, and Lachesis,
collectors and connoisseurs of the pit.

The souls of Kong Sheng's spectral shield-mates
drink deep, witness to Batiga-Shertko's
Narts and Uburs marching against Satan.

A sinkhole opens. The un-stench of cold
wind spreads, and out crawls Itztlacoliuhqui,
threading a deadly frost across the wold.

Out climb Jabru, Li Gong, Lethe, and Ji Bie,
who storm the distilleries to concoct
vile molotovs, mixing spirits with

corrosive poisons and venoms they brought
from the cauldrons of Kisin and Huang Xile.
On reeking plains, hooded Kumakatok

use such weapons to keep Luison at bay,
and stall Manes, in their gladiator's garb.
Protracted sieges of brimstone villas

stir up fallout, burnt blood, and
mustard gas, miasmas crafted by Dartsa-Naana,
where modern war's dead revisit their last

choking gasp. Mania and Mantus, clad
in robes of moth wing perfumed with
lotus oil, warn Liu Guangzhong of mad

Keres, and her fuming wrath. Hela, swift,
and Ishtar-Deela, charge at Lu Zhongce
and their colliding blades create a rift

in space, a door for Jihayeojanggun,
whose troops pour forth like wine. Macaria
challenges Keuthonymous, one-on-one;

Huang Shou, Mahakali, Mahākāla,
and Kherty start a betting pool. Ma
Zhong captures some Mani from Lampades,

whose banners of flame rain bitter ashes
upon Maximión as the former
retreats. San Pascualito gives the rash

host of his lord respite from marching's norm
to break their battle-fast. He bids Viduus
find fodder for a feast fit to adorn

the tables of soldiers allied with Muut.
Perched on a gypsum crystal, Melinoe,
Pana, and Santa Muerte pray for blood.

Viduus encounters Paowei and Minos,
and shares in their kill, an auroch of Hell
its ghostly flesh the prize of Odin's host.

The table of Mictecacihuatl,
and her counterpart, Mictlantecuhtil,
overflows with benthic fruits, creatures felled

by Namtar, Morana, and Rong Zhen: krill
the size of dogs, translucent sharks with eyes
bioluminous, mermaids caught by fell

fisherman of Ninsusinak's ilk, sly
eels with razor tails, octopi
and trilobites. Nga, Mors, and Nephthys

join the feast, while Viduus chats with Orphne.
With a wail of light, Osiris, crazed,
enters the hall. Tables fly, and the god

demands a portion of the feast. Raised
from his chair, Ogbunabali cries "Strike!"
His allies, vicious kin from lands decayed,

brandish their weapons. Bands of Mormo shake
their spears, Nergal unsheathes his blade, Morta
points her staff, Peklenc prepares an earthquake,

Qebehsenuef roars, Nenia Dea howls,
Proserpina removes her crimson shawl.
The waters of Phlegethon stall, sour

waves pouring from the lips of Styx enthralled
by cauldrons boiling over, frothed with war.
Rhadamanthus grimaces at the hall

erupting in antics of wasteful gore.
San La Muerte sighs. Shi Tong, unphased,
rattles the shackles of his slaves. Though sore,

Seker departs Osiris, wings painted
by war, and calls for Sidapa, Shingon,
Shiwang, Supay, Soranus, and the raised

Shinigami of Censors Youqing (Song),
Guan Yu, and Wu Lun, "terrors of the damned."
Judge Zi He, braced for chaos, warns Xie Bian,

master of morning's rage, to raise his hands
in preparation for the brisk collapse
of what tenuous peace once held the lands

of death in check, each master of each batch
of souls content, each land and fortress walled
and quiet. Some sick spark or rabid match

struck by a foreign hand set underworlds
at odds, unleashed both gods and darkling thralls
in this abyssal civil war. Whiro

allies with Xargi and Xolotl; mass
genocides follow this triumvirate.
Xipe Totec joins the fray. Through a glass

of red mirrors, Thanatos contemplates
his move, while Yin Changsheng summons Zhou Qi.
Yusai outfits his ships with cannons caked

with ancient blood. Judges Wang Fu, Zhang Qi,
Yang Tong, Xue Zhong, Zhao Sheng, and Zhou Bi stayed
to meet Zhu Shun, and watch Yan Luo, the

emperor of a perished palisade:
Youdu. Under the subterranean moon,
Tien Yan offers secrets to Wuluwaid,

in trade for passage across Varuna's
tumbling waters. While bleeding armies roam,
the demoness Vanth soars, her torch in bloom,

seeking the entrance to the fabled home
of Tartarus's heart. Uacmitun
wakes in this tomb, stirred by a dream of stone

and tranquil waters, where the dead were one.
Tuchulcha and Tuoni storm the famed
spires of Wu Yan and Wang Tong's prison

to recruit inmates for berserk campaigns.
Tusok Sacha, Vichama, and Yao Quan
counterattack. Yeomra shrieks in vain.

Zhen Yan, Zhao He, Zhou Sheng, and one Zhang Heng,
bolster the bold assault with forces drawn
from Yum Kimil, Ta'xet, and Wang Yuanzhen.

Yama and Wang Yuan bless their able pawns,
Yamaduta make pacts with Xun Gongda,
Veles march through dark woods of lives bygone

and Almas follow suit, with trudging claws.
alchemists that kind Māra despises
slip past Ghamsilg and Melhun, to a raw

outcrop of stone. A cry from far outside
the fabled lands and afterlives conceived
by living minds echoes: a neural tide.

Scant parchments of bark, lost but still believed,
foretell the obsolescence of the gods
of mortal fears. The universe aggrieves

our pitiful mythologies Beyond
the borders of each comforting fiction
told by the ashes of a flame long gone,

embers writhe on, in dreams, their dark friction
tugging at the soul's eye. Alchemists pray
for portals. Dagon and Yog-Sothoth trick

each parched soul walking Nyarlathotep's way
and Azathoth beckons them toward the gate.

CRITICAL MASS

Under the stands of an abandoned racket court
in Stagg Field, the last worker of his squadron
pauses next to a mound of graphite blocks
encased in fresh-cut timber. Stray motes

of sawdust cling to his face. Poor lighting and
peeling paint plague his private stage. Younger
feet have scuffed away the coating of the
floorboards. Cobwebs colonize every corner.

He wipes his brow. His nostrils twitch
at the room's stench of pencils, instruments
he once sharpened en masse as a childhood
punishment. He hesitates, reaching for the light

switch. The basement room becomes his boyhood
bedroom, where in those dreaded moments before
fatigue overcomes fear, familiar pieces of furniture
would morph, their structures ruptured by shadow.

He flips the switch, and the sleeping reactor
becomes an undiscovered temple of Babylonian brick
tainted with the soot of burnt corpses; a shrine
of Aztec stone crusted with sacrificial blood;

a sunken mass of Egyptian granite stained
with squid ink; the sand-scarred ruins
of a Nubian temple, consecrated with fading ichor;
the pyre of a Viking warrior giant, petrified

in a mudslide during a Celtic counterattack;
a forgotten vault of Hell's military fortress
Pandemonium, assembled by demons from bricks
of compressed crematorium ash and sanguineous

mortar; an unnamed outcrop of suspiciously angular
stone on an Antarctic peak; a Mayan pyramid replete
with apocalyptic glyphs; the rusting rubble of Chernobyl;
the deadly debris of Fukushima; the remains

of a Japanese estate that endured unblemished
for centuries, until a wave of fizzing sprites of light,
jostled from their subatomic limbo, swept away
its careful order. After the success of Fermi's reactor,

the physicists involved added their signatures
to an empty bottle of chianti fiasco, and buried
the reactor's remains in a concrete sepulcher
beneath the gnarled trees of Red Gate Woods.

THURINGIA

Stay out of your bedroom. There is work
to be done. Dreams are just screaming

with pictures. The pure of mind do not
slumber while final tests unfold. There

are cattle waiting for merciful slaughter.
There are documents that verify lies,

but why mention them? Radiation
permeates the pages of every archive

like a virus. Why bother validating lost
truths, when no one will hear them?

Desperate officers of the Nazi regime
test a prototype weapon on unwitting

German citizens and prisoners of war.
There is a flash of false light, a burst

of photons that makes newspapers
readable in the dark, their propaganda

briefly photographed by subatomic
scattering. Had the light persisted, each

reader would have been deconstructed
down to a collection of sparkling bones.

TRINITY

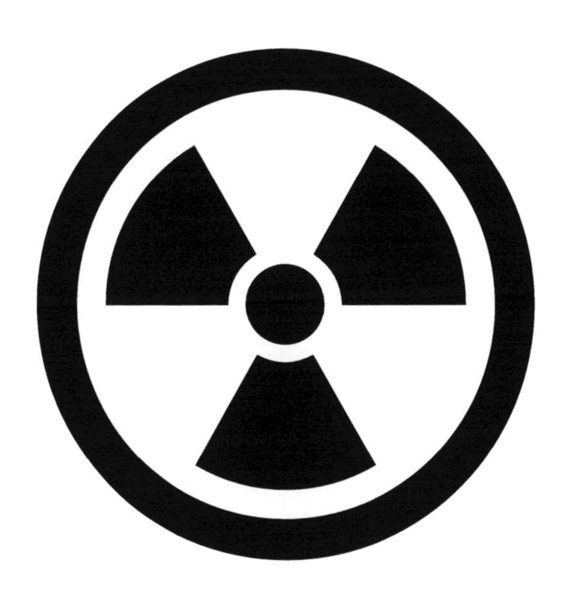

The night before, there was a lightning storm.
The bomb, hoisted to the top of an iron tower,
tempted the sky to ignite the kindling of hubris.

At dawn, the observers sheathed their eyes
behind protective glass. There was a wave
of thunder from the tower, and a scream of light.

Some experienced a lingering green glow behind their
eyes, as if their optic nerves had crystallized, fibers
of human cells transmuted into strands of candied opal.

In the aftermath's aurora, stray electrons
mingled with Earth's magnetosphere,
searing the sky with an orange tinge.

Metallic powders sprinkled from the
frothing cloud, whose blushing hive of
sparks hid microscopic cysts of isotopes.

Inside this cloud, resurrected warlords roared, their
fuming armies charging through geysers of infectious
grit, eager to infiltrate the future with their half-lives.

In this cloud, every atom sang in unison with rage.
Bunker designs and faux structures, crushed by later
blasts, would deflate into mounds of rubble like

the buildings of a child's toy village, kicked apart
by grinning siblings. The stunned team caught the sun
in the desert's throat, a decomposing strobe of red,

lashing their tissues with its tail of fire. The test site
echoed with the crack of Baphomet's whip, a gale
let loose from some lifeless plane of flame

that hickied their skin with an eerie kiss. Knives
of light sheered the air. Cirrus veils of red cracked
amid plumes of violet and blue.

The air folded apart
around the body of the cloud, a celestial wound
snowy with metallic ash.

A wave of seismic whiplash would describe, in passing,
how the bomb's threat would creep beneath all
discourse of war, promising a flood of fire

to drown each killing voice in the silence of its peace.
From the base of the bomb's chromatic cloud,
a skirt of noxious cream descended,

forming a conical gown,
in a tableau pirouette
of corrosive silk.

Above the gelatinous gaggle of dust (a brain
of frail, molten lobes), smoke rings blown from
Lucifer's lips billowed, spreading their toxic thrall.

Less than a month later, a bust of Mary was recovered
from a cathedral in Nagasaki, her eyes blackened
by a new species of light, the visage of the atom's age.

GHOSTS OF LOS ALAMOS

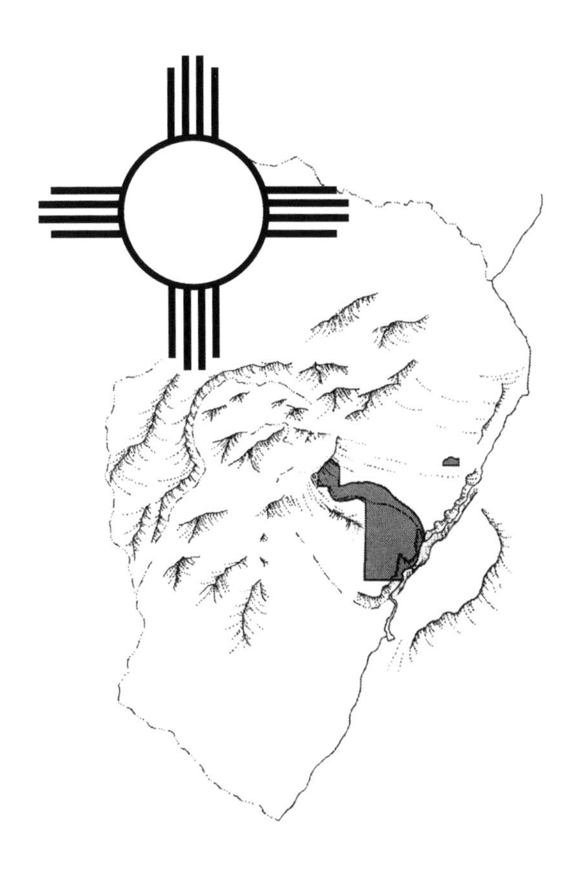

VALLES CALDERA

The pilot removes your blindfold. Mountainsides,
drowned in sunlight, make your eyes ache.

The terrain mutates from umber cliffs and dusty
shrubs into grasslands as gold as yellowcake.

Over the radio, generals scramble to snuff out
a global war, a manic brushfire fed by oiled steel.

You are their bubble in their think-tank,
a brute-forced cadre of tinkerers

commissioned to scour invisible realms
for elemental secrets.

Amid a ring of peaks in the Jemez range,
smoke seeps from fumaroles.

Sulfurous mud boils, the saliva
of a lock-jawed caldera.

On this volcanic plane, alchemic locksmiths
gather to forge the mould for a dark key.

As you fly over the region, shadow-blackened
craters at the caldera's heart form a pattern like

the pawprint of Cerberus. Do you dare
enter his den?

Brought into the fold, you'll posit ways in which the
West might outpace the Axis in coaxing misfit isotopes

towards a controlled subatomic civil war, might goad
atoms to renounce their fragile yet familiar molds,

aid in perfecting a process of atomic distillation
using flurries of precise incisions applied to

innocent nuclei, levied upon swarms of
unseeable, humming yolks: a fatal surgery.

Eclectic clouds of electrons struggle to cling to their
native bodies of protons and neutrons. Technicians

sweat at metallurgical looms, entombing and
exhuming volatile rods encased in concrete.

These graphite mausoleums enable the pursuit
of destructive constraints.

Encouraged to commit sedition, stray neutrons
knock tight-knit units of their brethren apart, as if

made jealous of such cohesion. Rogue particles
clatter through stable collectives, shattering weak points,

breaking symmetries, disrupting stabilities, spoiling
cohesive bonds, rupturing and rending. Lo and behold:

Humanity has taught the building blocks of matter to
adhere to our particular designs of war. And yet,

not long after each radiant outburst, the enduring grace
of radiation laces untouched glades, brittling floral tissues.

Subtle waves infest charted seas with martytred particles.
Contaminants of varying gradation settle

in colonial abodes. Choirs of lyres bloom in unison, their punitive
tones lamenting plutonium's decay.

Diana sighs for her lost prey caught
in the decay of Apollo's latest melody.

THE DEMON CORE

for Aaron Tucker & Michael Lista

Past wind-combed sagebrush and sedge grass
the top secret hovel lies. Makeshift parties in this
military nowhere, fueled by punch spiked with lab alcohol,
offset late-night sessions of chalkboard chatter.

Talk of death toll estimates and explosive yields
mingles with the distant cries of coyotes fleeing
exterminators. In the wet season, an incessant murk
of mud swallows sleek automobiles, and

hastily dug wells often cause bathroom taps to
spill forth earthworms. What more fitting a
terrestrial Hell could these idealistic
students have found themselves invited to?

Technical assistants heave a rectangular metal case
into the newly-established lab, a glorified cabin,
where scintillation counters will measure
the tipping point of an orb of plutonium-gallium alloy

surrounded by neutron-reflective bricks of
tungsten-carbide, the whole assembly glistering
like a forbidden treasure, a regal orb
heavy with curses.

Undisturbed, this deadly sphere retains the
chill charisma of a ball of ice incubated in a
copper cylinder by a skilled bartender, in order
to tweak the temperature of a chic drink.

The wrath of this nuclear artefact catches
two students in a glare of blue light,
unleashing upon each of them
several lifetimes of sunlight

in a terrestrial flare, like a swollen star
blowing apart the
living formulae of their
delicate cells.

After this disaster the orb will lurk, inert,
amidst the debris of its making, waiting to be unleashed
at Bikini Atoll, where it will boil the sea
into a flowering tower of froth and pestilent seeds.

Science will learn soon afterward that,
as the dinosaurs began to die, gentle breezes
carried pollen through the oxygen rich air
ushering in the appearance of the orchids.

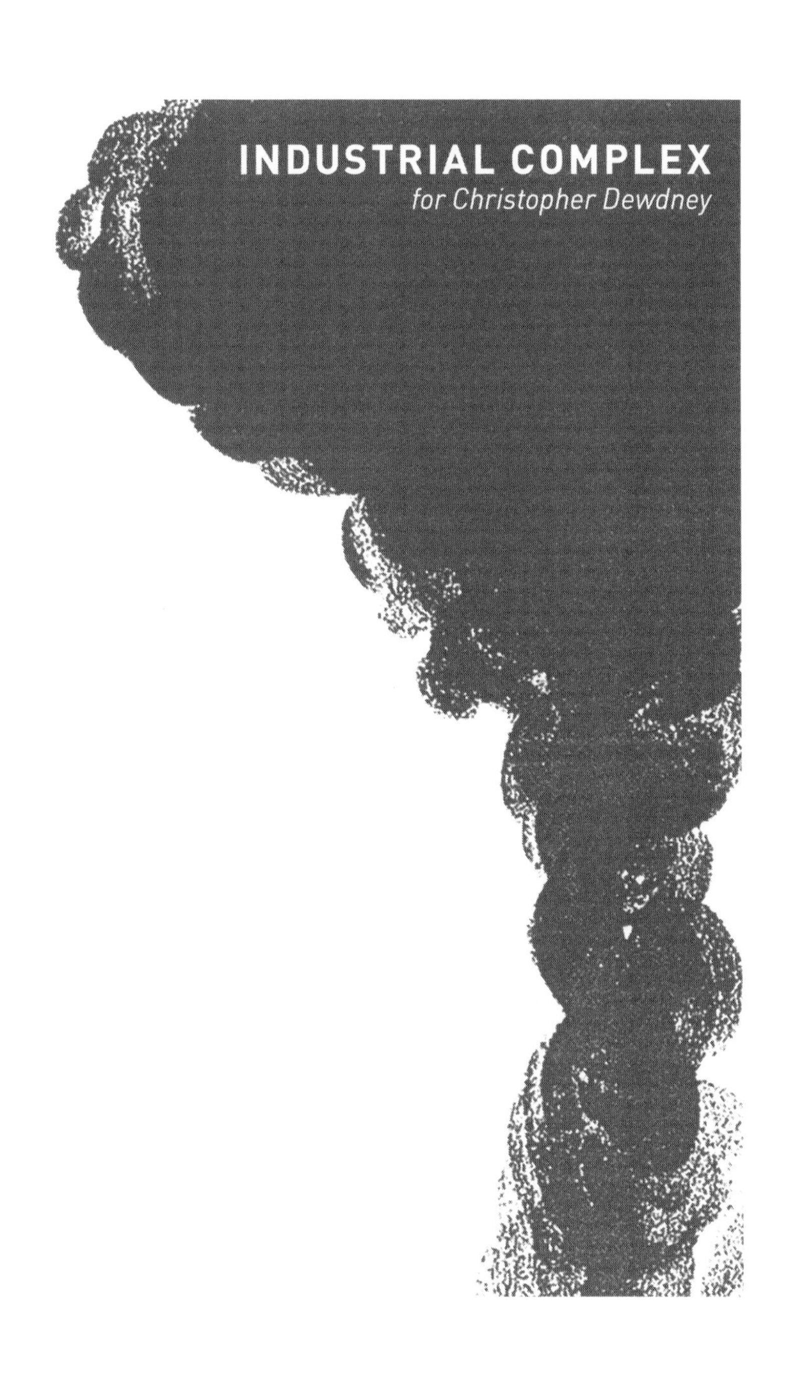

INDUSTRIAL COMPLEX

for Christopher Dewdney

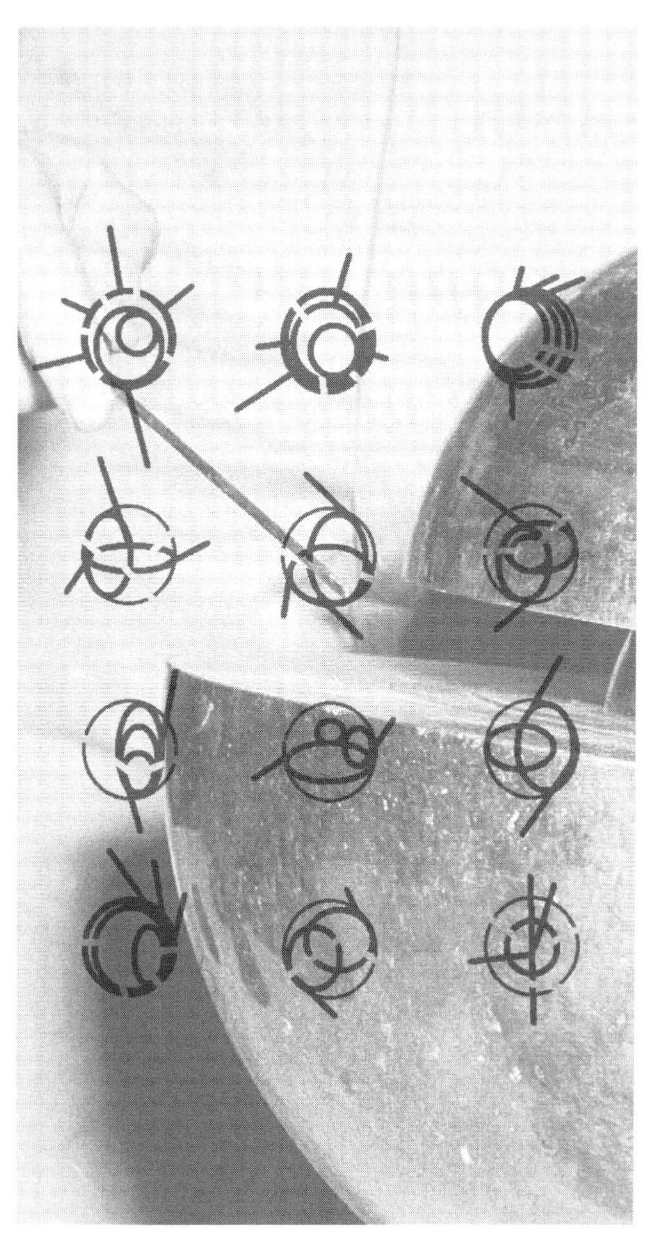

THE ENCORE FOR DR. FAUSTUS

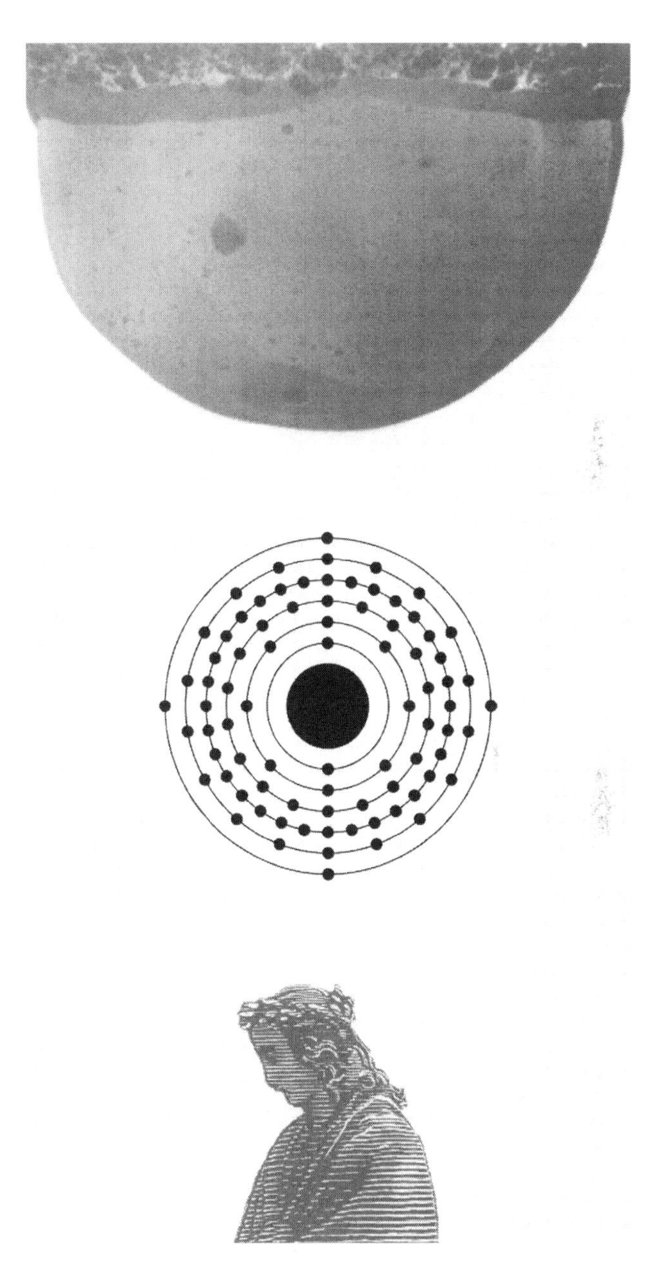

THE DAWN OF OUR FALL

THE SLEEP OF REASON

THE COURT OF THE FATES

MILITARY INCIDENTS

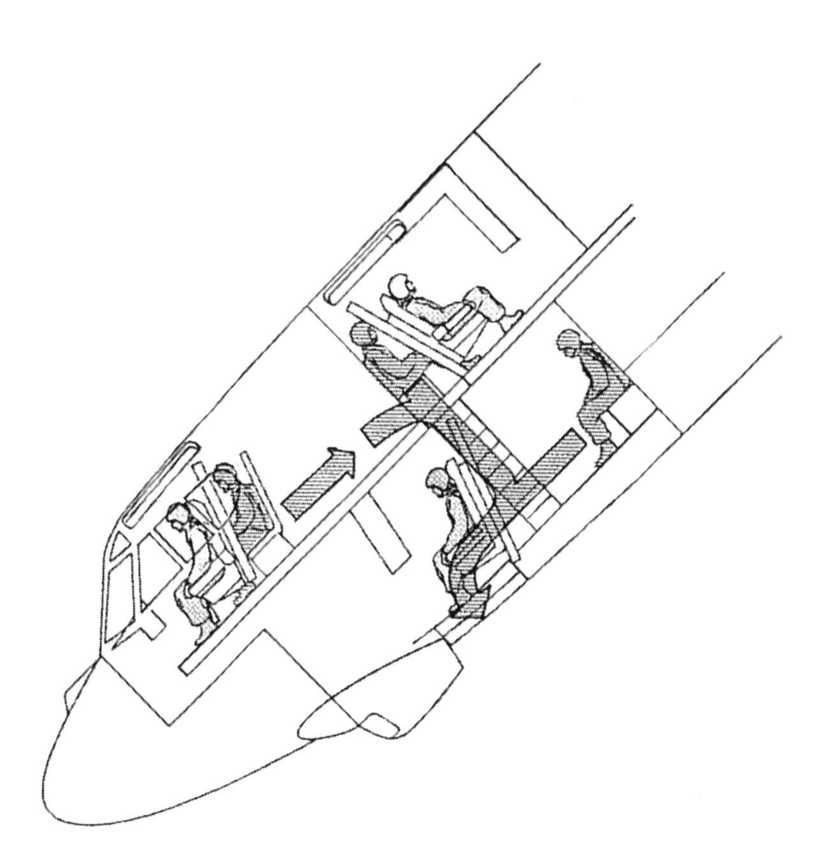

DULL SWORDS

Flint tools fracture and snap
in cro-magnon skirmishes.

A pig iron blade fails
in a shower of shards.

The blade of a scimitar
loosens in its worn hilt.

A blunt scythe fails
to fell a bunch of wheat.

A pair of butterfly knives succumb
to rust in the unmarked tomb of an assassin.

The edge of a glaive, caught in the crevice between
two cobblestones, splinters along its pole.

An apprentice's error ruins
A katana blade, incurring
The forgemaster's discipline, which
He exacts with a bamboo staff.

A torturer bends the blade of his prisoner's prized
rapier, in a vice, right before the swordsman's eyes.

A machete saves an explorer, slicing clean through
an attacking anaconda, only to corrode
in the serpent's digestive juices.

Rome burns.
A gladius melts in the debris of
an exploded kiln.

An avalanche in the Swiss alps buries
a custom zweihander, along with its wielder.

At the bottom
of the Atlantic Ocean
slowly swallows
a pirate's cutlass.

A shipment of falchions tumbles down
a gulch, as does its caravan.

A pharaoh's khopesh cracks, improperly
packed for shipment
to a museum.

A tarnished katar fails to penetrate
an enemy's armour.

An oxidized kris
fuses to its sheath.

A bolo breaks.

A pistol-sword
misfires.

BROKEN ARROWS

February, 1950:

> Ice collects on the air intake of a British bomber equipped with a mark IV nuclear bomb. The plane jettisons its cargo, which explodes over Alaska's Inside Passage. Canadian authorities are not told what kind of ordinance the bomber was carrying.

March, 1956:

> A Boeing stratojet leaves MacDill airforce base with two containers of weapons-grade nuclear material. While the material onboard could not have caused a thermonuclear explosion, neither the jet, nor its crash site or debris, are ever found.

February, 1958:

> A fighter plane hits a B-47 carrying a mark XIV nuclear bomb, which is jettisoned. The bomb, still lost in the Wassaw Sound, should have contained a dummy core. Testimony from former defense secretary W. J. Howard claims otherwise.

March, 1958:

A Boeing stratojet leaves Hunter airforce base, carrying a coreless mark VI nuclear bomb. The pin that locked the bomb's harness fails to engage, and the bomb falls on a playhouse, nearly killing two nearby children, and creating a 70-foot crater.

January, 1961:

A B-52 stratofortress, with two mark XXXIX nuclear bombs, crashes due to a fuel leak. The bombs land without detonating. Only one of the four safeguards for each bomb remains intact. Much of the nuclear material from one bomb is not recoverable.

March, 1961:

A B-52 stratofortress carrying several nuclear weapons runs out of fuel when its crew (prescribed amphetamines to combat fatigue), refuses an emergency refuelling. The aircraft crashes. None of the nuclear weapons detonate due to failsafes.

January, 1964:

The vertical stabilizer of a B-52D with two nuclear weapons snaps off in a blizzard. The craft crashes in an Elbow Mountain meadow. Three crewmen die. A team recovers the bombs. They discover the ship's navigator and tail gunner died of exposure.

December, 1964:

A B-58, trying to take off from Bunker Hill airforce base, skids off of an icy runway, colliding with an electrical box. The aircraft catches fire. The nuclear weapons onboard are scorched. Radioactive contaminants are confined to the site and removed.

December, 1965:

An A-4E skyhawk falls into the sea from an aircraft carrier, during a training exercise in Subic Bay. The plane, along with its pilot and the B43 nuclear bomb aboard, are never found. The pentagon releases no information on the aircraft until 1989.

January, 1966:

A B-52G bomber collides with a KC-135 tanker while refuelling in mid-air. The tanker explodes. The B-52G crashes. It's cargo, four Mk28 hydrogen bombs, stays onboard. Explosives in two bombs detonate, littering the crash site with plutonium.

January, 1968:

A cabin fire in a B-52 bomber prompts its crew to leap to safety. The B-52 crashes in Greenland's North Star Bay. Explosives detonate in all four hydrogen bombs. A clean-up is performed, but much of the radioactive material is not recovered.

September, 1980:

Working maintenance in a silo in Arkansas, a technician pierces the fuel tank of a Titan-II nuclear missile. The silo explodes. Explosives in the missile's warhead detonate. The warhead's failsafes prevent a loss of radioactive material.

BENT SPEAR

August, 2007:

> At Minot airforce base in North Dakota, personnel mount six AGM-129 nuclear missiles to a B-52H. The warheads should have been removed from the missiles, but protocol was not followed. Four commanders and several personnel were disciplined.

EMPTY QUIVERS

An amateur fletcher
fumbles, nicking his fingertip on the point
of an obsidian arrowhead.

A batch of bows, their wood mistreated,
fails to bend.

The treated leather
of an archer's quiver
rots in a family crypt.

Straw targets,
packed too loosely,
allow arrows through them,
injuring a passing squire.

In Rome, the head of a
legionnaire's spear detaches mid-air.

The firing mechanism of a crossbow
jams.

Harpoons bounce off of the
blubber of a whale.

Ballistae burn.

A slingshot snaps.

A ninja drops a throwing star,
blowing her cover.

An English longbow, enshrined in a private
collection, turns to charcoal as the manor burns
in a fire set by a neighbourhood pyromaniac.

Throwing knives, dropped on the battlefield,
crumble centuries later under
the blade of a farmer's plow.

A weathered stone, once thrown from
an ancient sling, becomes the centerpiece
of a CEO's zen garden.

A cannonball barrels through a
suburban neighbourhood, escaping
an experiment on a nearby firing range.

A musket ball explodes in its barrel,
embedding shrapnel in the eyes of a
Minuteman.

A minigun overheats, its barrel
glowing molten red.

A luger jams, foiling a
suicide attempt.

FADED GIANTS

With one fabled swing of a stone
from a homemade sling, David blinds Goliath.

Moses slays Og, King of Bashan.

Nephilim, angelic half-breeds who
escaped Noah's flood, wither away
in the windswept corners of
desert kingdoms:

Anakites, Emites, Amorites,
and Rephaites. Their bones
evaporate, reclaimed by heaven.

Gogmagog flaunts his
bastardized name in the
hills of Alvion, until
Corineus heaves him
from a cliff.

Joshua banishes Anakim.

Uj-ibn-Anaq, who stood
knee-deep in the open ocean,
slips and drowns
in the Mariana Trench.

The Ana, artisans of
human lifespans, succumb
to cancer.

Daityas fall before the gods
they fought in jealousy.

The Si-Te-Cah experience
total societal collapse,
driven to cannibalism
by human foes.

A snakebite finishes Orestus.

The body of Ajax resurfaces,
each kneecap
the width of a discus.

Odysseus blinds the Cyclops.

Heracles defeats Antaios.

An oil spill smothers the last
of the fifty-headed Hyperboreans.

The Laestrygonians commit
ritual suicide.

Odin and his kin
kill Ymir, who was
born from droplet of
meltwater hanging from
a poisonous icicle.

Kerlig the hag
laughs herself to death.

Fafnir morphs into
a hoarding dragon,
a ripe target for Sigurd.

Freyja retreats to avoid
a celestial civil war.

Neringa's heart gives out
in a contest with a
dragon-fighting Isopolini.

A mortar shell blows
the Bergmönch
to bits.

Antero Vipunen,
the giant shaman,
relinquishes the
three words of a
powerful incantation
to the god-hero
Väinämöinen,
after the hero opens
his grave and
skewers him
with stakes.

NUCFLASH

because one sun
was not enough

RAIN OF RUIN

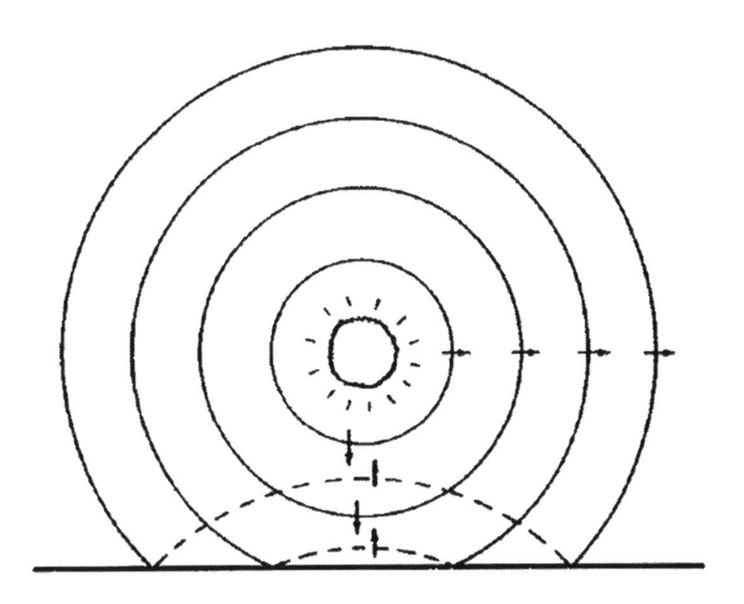

CLEAR SKIES

We

belong to the

wind

of a late

dawn

a Bomb

touches

her

target

and

We turn

to

smoke

groping for words

we

live

on

TESTIMONY

for Mariko Nagai

There was a flash, a thermal lance of
magnesium. White clouds spread out from
the glare, a morning glory blooming
in the sky. There was a blast

of steam. I felt weightless, as if I were an
astronaut. I was blown into another
room. When I regained consciousness, I
found myself in the dark. I was caught

under something. I fought to escape.
I thought maybe I was having
some kind of nightmare. The dust was
rising. Something gritty

entered my mouth. My clothes had
turned to rags. Thinking that my
house had been hit by a
bomb, I removed the red soil

and the roof tiles covering me. It was
as if a box of matches had been struck by
a hammer. It smelled like a
volcano. I heard people crying for

help and for their mothers. The cries were
coming from underground.
There was a sheet of flames in front of
me. A burning whirlpool approached from

the south. A tornado of fire, spread over
the width of the street, approached from
Ote-machi, scorching my ear and my leg. I
didn't notice these

burns until later. What impressed me strongly
was a five-or-six-year-old boy with his
right leg cut at the thigh.
He was hopping on his left foot

to cross over a bridge. After a while, it
began to rain. The rain was black. The
fire and the smoke had made me
thirsty and there was nothing

to drink. I opened my mouth and
turned my face to the sky. Maybe I
didn't catch enough rain, but I still felt
thirsty. The fire didn't subside.

The river was filled with dead people and
with survivors who came there to seek
water. I could not see
the surface of the river. I took care

of the people around me by using the
clothes of dead people as bandages.
Hiroshima was covered with only three
colours: red, black, and brown.

The fingertips of corpses caught fire and
the fire gradually spread over the
bodies. A light gray liquid dripped down their
hands, scorching the skin.

I saw the father of a neighbouring
family standing almost naked. His skin was
peeling off all over his body
and was hanging from his fingertips.

I tried to talk to him but he was too
exhausted to reply. After the bombing, I felt
paralyzed whenever I saw the sparks made by
trains or lightning. At home,

I could not sit beside the windows because
I had seen so many people wounded by
pieces of glass. I sat with the wall behind me
for ten years.

OPERATION EPSILON

No one has any money in Germany.

I would have no pangs of conscience
making neutron sources for
the Americans.

They have money and,
in consequence,
they have time.

We will pay for being here.

The day before I went away, I
said to my wife,
“I suggest we commit suicide”.

It is the future that worries me.

I should like to work on the
uranium engine.
I should like to work on
cosmic rays.

Once, I wanted to
suggest that all uranium
be sunk to the bottom
of the sea.

I wonder, are there
microphones
installed here?

CONTAMINATION

for Valery Legasovi

CHRISTMAS ISLAND

for Ted Blackwell

At Christmas Island
I saw three atom bombs
being dropped.

They gave us photos
of the tests, which I
have somewhere.

We had a routine.

Some would wear
anti-flash goggles
and sunscreen.

You had to turn your
back, kneel down, and
hunch over.

It was like when you're a
child and you press a
flashlight to your
skin, you could see a

red glow,
and the shadow of
your bones.

There wasn't a mushroom cloud.

It was stranger than that, this
glowing mass climbing
through the air like a
ball of serpents.

It burned the sky
for three days,

blinding
all the birds
on the island.

It was only when their
chicks hatched
that the birds
could see again.

PLUTONIUM VALLEY

for Craig Dworkin

Alkali accrues in basins of abandoned badlands.

Pools of brine give way to splitting planes
of hardened clay, replete with radial cracks.

Brittle crystals beard the
hardpan lakebeds.

Legions of salt carve sharp but fragile letters
into the weary earth.

Ancient colonies of dormant halophiles

huddle in the drained tributaries of Styx,
alien archaea thriving on thin films of saline.

Evaporated water leaves behind
minerals too heavy to
join the clouds.

Golems hide under
the rubble of barren hillsides, glaring at
military personnel from the mirrors
of their dreams, sand looking
through sand.

The troops build follies in this wasteland,
each phantom dwelling an offering
crafted for each bulb of flame they will unleash.

Tall flowers of fluorescent smog
seize the military onlookers, poorly braced
for the shock of such awe.

Those still standing fall to their knees,
as mach fronts shake the ground.

Branches of thunder spread,
leaving lines of Joshua trees in flames,
like the climax of a Pagan rite.

Observers in Vegas
attend bomb parties at dawn,
on hotel rooftops,
breakfasting
in the distant glow
of hydrogen fusion.

And inside, in the bars, crystal glasses
sing with the resonance
of every thundering blast.

THE EAST URAL RESERVE

Pine needle fall from their trees
in radial shadows, a slow green snow
acidifying the soil, curating beds of
alkaliphilic flowers.

legions of heavy metals, actinides
and lanthanides, bloated with
excess energy, force nature to
accommodate
the atomic weight
of human folly.

The strain of such decay
remains, recorded

in the ringed stumps of trees
felled in the wake of a disease, their
toxic bodies and stiff limbs yet lively
with a mutagenic heat.

The more deftly a poison hides,
the greater burden of fear it creates.

A sickness weighs on this innocent landscape.

Radiation falls in fistfuls of hail,
pummeling the plains of Siberia,
hammering untrodden meadows
briefly freed from human occupation,
but not from human impact.

The obituary of humanity
resides, not only in the cargo of our
fragile probes, but in the spikes
of fallout we have written in the Earth,
like atomic specimens preserved
in slides of strata.

For decades, cleanup efforts fail to address
the weight of grains of sunlight, stirred up
from remote barrows dug to entomb
caches of spent nuclear fuel.

This insidious detritus endures,
woven into all forms of weather.

A stray snowflake, bearing
a frozen grain of oblivion's brine,
dissolves on a child's tongue.

THE ARGONNE INCIDENT

The coffins of the three dead men were lined
with lead before their bodies were allowed
to attend their funerals.

The earth was briefly spared the burden
of their unnatural decay.

The ancient practice of Charon's Obol
demands that a coin be placed over
the mouth of the deceased, as fare
to ferry the tired soul across the river Styx.

Can lead, pressed into the shape of a coffin,
still serve as valid payment? Or will a curse,
composed of livid atoms, follow each
exhausted spirit
over turbulent waters, to dark beaches
choked with the weary dead?

The simplest flicker of a single,
inattentive mind, caused by a stray
thought, or rogue
memory, wandering through the brain,
could re-open the door for death.

A bar of music from a childhood song,
the ghost of a favourite ice cream flavour,
the notion of some ridiculous love
that never was
and never
would have been,

could induce criticality, like a control
rod withdrawn too far
could give rise to a wraith of steam,
a "water hammer,"
which it did,
hurling a twenty-six thousand pound reactor
two meters into the air.

A shield plug
pinned one man to the ceiling,
acting as a searing spear,
a deadly javelin of Belial.

Various investigators sifted through logbooks
with their best questions prepared:

Did corrosion or wear cause the fatal rod to stick?
Did the sticking prompt too harsh a pull?
Was sabotage a possibility, born from some hatred
shared between the three men?
Something seething underneath the surface?
Perhaps a buried affair or a grievous injury was left
to enrich itself, uncontrolled, until
it went supercritical, erupting in that
single motion, a freak murder-suicide?

No. The dreaded conclusion admits
the cause was simply
a mistake
free of the twin comforts
of intent and cause.

THE HUMAN FACTOR

A deluge of corium sludge
eats through steel, lead, and concrete,
with its fleet of subatomic jaws,
chewing and subsuming
each substance it touches.

Frantic employees fiddle
with inoperable equipment,
relying on inaccurate dials,
as others compute the radius
of the potential blast.

Citizens are herded from their homes
under the drawl of sirens and the
clarions of alarms.

Megaphones blare,
protestors resist,
news anchors
swarm to the story.

Less than two weeks prior
to the partial meltdown,
the film *The China Syndrome*
depicts an identical event,
and even mentions Pennsylvania,
the unfortunate state where
Three Mile Island resides.

A single, ambiguous indicator light
and a poorly-placed pressure gauge
set a cycle of assumptions in motion.

Each operator remains soundly trapped
by seemingly infallible tools, repeating
ineffective procedures with increasing
frustration.

The reactor only cooled
after a fresh team arrived
to replace the old.

THE WIDOWMAKER

for the victims of the K-19 disaster

Draped in red and gold,
the Soviet submarine K-19
awaits its christening.

Russian officials ignore
tradition, and a man, rather than a woman,
releases a champagne bottle tied to the end

of a pendulum of rope. The bottle
fails to shatter against
the submarine's steel hull,

a bad omen, yet insufficient
for the horrors ahead.
No amount of prayer will heal

those blinded by the radiation leaking
from the submarine's
shoddy reactor, a pressurized beast

repaired multiple times, under duress,
miles below subarctic waves.
In such conditions, the mind becomes

nothing more than a cerebral oven
of radiation, barely able to hold the image
of a lover in its frail
and failing tissues.

THE ELEPHANT'S FOOT

for Joyelle McSweeney

Smog seeps from the somber concrete
factories of industrial hamlets, the workers' hovels
bordered by the shoals of ghostly lakes.

Dubious institutions offer guided tours
of the ghost-city of Pripyat and the exclusion zone
surrounding the Chernobyl plant.

Embark on this excursion, but remember to burn
the clothes you wore when the tour ends. You have
no way of knowing what has nestled in your fabric.

As you pass the first military checkpoint, your guide
reminds you that to stray too far from the tour group
incurs a penalty that he refuses to describe.

You pass through
a guarded gate
and the spell begins.

The narrow road decays. Rabbits scavenge
among the wily weeds springing from tired asphalt,
acting as unwitting sponges for belligerent isotopes.

Arthritic trees with prematurely-crisping leaves leer
at human caravans, as if each molecule of mutant
chlorophyll concealed
a bloodshot eye.

Spears of marching grass patrol the road,
prodding beneath thickets of luminous trees
and alongside swamps rich with atomic rot.

What sap sleeps in the trunks of diseased woods?
What honey waits in globulant hives? What fungi
runs amok among networks of roots?

Asphalt roads dissolve into lagoons
of black shards, dark as the damning photos
radiation rendered black, in auto-censorship.

Vintage maps of the Chernobyl plant remain
dangerous to handle, still tingling with the
otherworldly heat created when water
molecules broke apart in the screeching core.

As you travel, your thoughts are haunted
by the specters of long-dead firefighters
at the site of the disaster, trying in vain
to quell a stubborn flame.

The caravan continues, passing abandoned
high rises, shawls of tattered curtains fluttering
from broken windows. Inside one such building,
your guide explains how intrusive photographers

have repositioned debris in order to compose
more poignant snapshots of the tragedy,
unearthing objects that might have remained
buried, or placing fraying children's toys on beds.

The weather changes as you pass the edge
of the vigilant lid built to entomb the first sarcophagus
of concrete to cover the reactor, the tut tut of rain
on the corrugated roof of the giant lid reminds you
of a funeral sprinkled with the cliché of rain.

As you pass a metallic arch, the dead reactor
rises from behind the black water of rolls of
undeveloped film, a cathedral of rubble under
a thinning moon.

Your guide points out that, in order
to prevent the contamination of an aquifer
that leads to the Black Sea, conscripted workers
dug a tunnel under the ruptured core

as its slobbering magma crept toward them,
a national disgrace hidden at the expense
of subjects of the state. Warped by their task,
the excavators stumbled back into the light,
each sore from bearing a cocktail of mutagens.

Today, an accidental taste of contaminated
sand on your tongue could trim a decade
or two from your lifespan.

Then, the makeshift miners dug without protective
gear, shirtless and maskless. They drank water from
open bottles, in tunnels where fans were banned
to stop the stirring-up of fatal dust. On the roof,
graphite debris stalled clean-up operations.

When robots failed to clear the detritus,
humans served as ideal replacements.

Equipped with shovels and lead vests, these
"bio-robots" were compensated for this
required risk with a few rubles
and some vodka.

You picture their winter of ash,
hoarfrost on scored mortar,
fields of falling debris, the sun a
broken machine, its last
incendiary breath
flushing the sky with
a lover's blush.

Back at your hotel, you lie,
suspended between guilt and
serenity.

Noxious stalks of moss, lying safely
many kilometers away,
puff out their deadly incense.

You have brushed
against the coat of death.
A rogue species
of iodine, so close to the one
your parents used to
clean your wounds with,
has crippled generations.

RISING WATER

The elderly bow to the young, resolving
to preserve the growth of trees that they
will never see grow tall enough to offer shade.

They journey to where Geiger counters peak,
to where the air twitches with subtle corruption
and the sea whines with seeping confusion.

Birds worry in their nests
while the restless moon tugs
at the bedsheets of the Earth.

In the most contaminated areas, volunteers
remain serene, neither rushing nor hesitating,
nobly weathering this tender storm of time.

Some wear humble smiles. Others,
apprehensive, foster tight-lipped stares
while facing down a triple-meltdown.

Escaped particulates slip into the blood
of a populace, ransacking their cells, creating
mutations that take years to incubate.

Dead wasps, embalmed with pollen, fall
from clogged eaves troughs, their limp bodies
bright as warning flags.

DOOMSDAY MACHINES

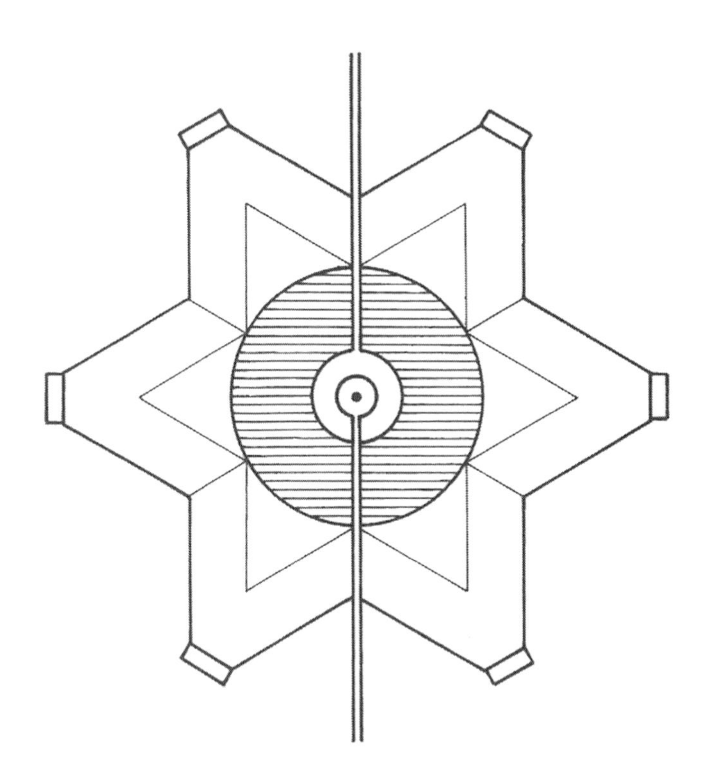

SALTED BOMBS

for Sandy Pool

Scribes struggle to recount the advent
of doomsday. Scavengers scrape radium
from watch dials, extract cesium
from stolen lab samples, raid

factories that manufacture smoke detectors
for their americium, and dismantle
glowing rifle sights for thorium.
Cell by cell, each dream degrades into

a fractured simulacrum. The vigilant
heave the remnants of their blasted hopes
into shallow ditches, where the pieces
burn with supernatural fire.

Invisible flames devour
their devoted hands. The dour immortality
of death endures, amid retaliating grass, within
the changeless grins of irradiated billboards,

and in each cringe of failing stone.
Ruins, rich in rebar, rise from pools
of scrap, rusted nests cloaked in hazes
of livid decay. Every well

brims with water sore with iodine.
The faces of drowned warriors arise,
caught in the eerie contours
of the carapaces of Japanese spider crabs.

In the expanding night, Artemis waits
to soothe each fevered soul, to strike
with her winter's graceful whip
of frost, a mother's damp cloth

applied to a burning brow, the gesture
sudden, yet gentle. When her time arrives,
she will salve each wounded, living text,
reset all clocks, fill in each fissure split

and splitting, reconcile each
division, wipe aside each
unresolved conflict, each
unrevised error, each

microscopic stillbirth, each
contorted nuclei. The forests
can afford to wait, but can enough
ragtag bands of families endure,

scattered across depopulated
hillsides, sheltered from black rain
by the arms of abstract cenotaphs?

They find refuge
In the gaping mouths
of ruins, on whose tongues
they parlay
with starvation.

ANTIMATTER

Ram, maim.
Ream, mar, taint, tear.

Man, tame
matter (a meaner titan).

Tan marine,
re-train an inmate.

Team, emit
trite mania. Maintain it.

Rain, tint
a marina. Tire an airman.

Art, imitate.
Eat time. Entertain me.

Mare, tatter
an era in mire and rime. Amen.

STRANGELET

Stags rest
at garnet altars.

Alert anglers
snag gentle eels.

Slant lenses
target star nets.

States regret
a Tsar's steel glare.

Stern lasers
test astral sleet.

Art tangles
strange letters.

Ants stare
at a greater stage.

RED MERCURY

Beet-red nectar, mixture
of liquor and Earth's blood,
Ba'al's concoction, taught

to a reverent few, long-sought
alchemic miracle, spritz of gold,
brief bath of basalt, din in mud

performed for beasts of deep
delights, eaters of clastic
delicacies, monsters born from

mournful moulds, aficianados
of repressed reactions captured
by rare intoxicants, harbourers

of lost tastes, the fell prey of time.
Even nuclear materials bow
to curious words. Whatever

fluids rest in the archival vials
of subterranean labs, whatever
insights writhe within suitcases

passed between tyrants, the most
volatile element, poetry,
preserves itself.

DESTROYERS OF WORLDS

for Moez Surani

UNITED STATES

Current Stockpile: 6,800

Alberta
Crossroads
Sandstone
Ranger
Greenhouse
Buster-Jangle
Tumbler-Snapper
Ivy
Upshot-Knothole
Castle
Teapot
Wigwam
Project 56
Redwing Project 57
Plumbbob
Project 58
Project 58A
Hardtack I
Argus
Hardtack II
Nougat
Sunbeam
Dominic
Fishbowl
Storax
Roller Coaster
Niblick

Whetstone
Flintlock
Latchkey
Crosstie
Bowline
Mandrel
Emery
Grommet
Toggle
Arbor
Bedrock
Anvil
Fulcrum
Cresset
Quicksilver
Tinderbox
Guardian
Praetorian
Phalanx
Fusileer
Grenadier
Charioteer
Musketeer
Touchstone
Cornerstone
Aqueduct
Sculpin
Julin

RUSSIA

Current Stockpile: 7,000

First Lightning
Joe 2
Joe 3
RDS-4
RDS-5
Joe 4
RDS-9
RDS-37
Tsar Bomba
Chagan

UNITED KINGDOM

Current Stockpile: 215

Hurricane
Totem
Mosaic
Buffalo
Antler
Grapple
Vixen

FRANCE

Current Stockpile: 300

Gerboise
Blue
Agathe
Aldébaran
Canopus
Achille
Xouthos

CHINA

Current Stockpile: 260

Project 596
CHIC-2
CHIC-3
CHIC-4
CHIC-5
CHIC-6
CHIC-7
CHIC-8
CHIC-9
CHIC-10
CHIC-11
CHIC-12
CHIC-13
CHIC-14
CHIC-15
CHIC-16
CHIC-17
CHIC-18
CHIC-19
CHIC-20
CHIC-21
CHIC-22
CHIC-23
CHIC-24
CHIC-25
CHIC-26 (aborted)
CHIC-27
CHIC-28

CHIC-29
CHIC-30
CHIC-31
CHIC-32
CHIC-33
CHIC-34
CHIC-35
CHIC-36
CHIC-37
CHIC-38
CHIC-39
CHIC-40
CHIC-41
CHIC-42
CHIC-43
CHIC-44-1
CHIC-44-2
CHIC-45

INDIA

Current Stockpile: 120

Smiling Buddha
Shakti-1-1
Shakti-1-2
Shakti-1-3
Shakti-2-1
Shakti-2-2
Shakti-2-3 (cancelled)

PAKISTAN

Current Stockpile: 130

Chagai 1-1
Chagai 1-2
Chagai 1-3
Chagai 1-4
Chagai 1-5

NORTH KOREA

Current Stockpile: ~15

1
2
3
4
5
6

FALLOUT

The Manhattan Project draws its theoretical basis, in part, from Joyelle McSweeney's *The Necropastoral: Poetry, Media, Occults.*[1] McSweeney characterizes the "necropastoral" as a liminal zone, containing "the manifestations of infectiousness, anxiety, and contagion . . . present in the hygienic borders of the classical pastoral . . . [a] location [that] stages strange meetings."[2] The necropastoral explodes binary distinctions like urban vs. pastoral, or natural vs. manmade. Instead, human influence, from our pollutants to our reshaping of the environment, and nature's re-infiltration of previously human-occupied areas constitute a kind of exchange—an osmosis of poisons, pests, roots, and chemicals across various membranes. The concept of the necropastoral urges us to consider the idea that what we think of as "natural" has always already been contaminated by human activities, while also pointing to the pervasive and vengeful infiltration of "natural" forces in so-called "urban" environments. *The Manhattan Project* explores the aesthetic milieu of the necropastoral in order to revive the tonal and imagistic elements of both the pastoral form and the lyric form for the post-nuclear age. In doing so, the book hopes to show that these forms cannot be resurrected as they once were. Instead, what rises from their lead coffins are far stranger beasts.

The Manhattan Project's[3] opening section, the poem "The Atoms We Cleave"[4] (its title a phonic imitation of the phrase "Adam and Eve"), describes a "tree of death" as an emblem of pollution, the antithesis of the biblical "tree of life." The places of the poem, affected to varying degrees by human activities, demonstrate that polluted and abandoned zones of human activity at once infest, and are infested by, nonhuman forces. A retinue of nymphs mourns the deaths of those claimed by disasters such as the Chernobyl meltdown, as they wander through scenery reminiscent of the locations used by filmmaker Andrei Tarkovsky in his 1979 film *Stalker.*

The next section, entitled "The Arms Race,"[5] spans the discovery and militarization of nuclear energy, beginning with fission reactions that took place in Oklo, Gabon, approximately 1.7 billion years ago.

The poem "Below Oklo" argues that "[n]ature was never innocent" by revealing not only that human beings were not the cause of the first fission reactions on planet Earth, but also implying that the discovery of this energy source was inevitable, given our ever growing appetite for energy.

The next poem, "Radioactivity," chronicles Marie Curie's discovery of radioactive elements—the pursuit of which led to her death. The poem ends by lamenting the seeming inability of language to convey emotive urgency, and thus to do more than occupy the "chemistry"—the ink and paper—that incarnate written language.

The poem "The World Set Free" dramatizes Leo Szilard and Otto Hahn's well-intentioned contributions to atomic science, contributions that were ultimately weaponized by military forces.

"Ideal Isotopes—Prelude" provides a brief description of radioactive isotopes Uranium-235[6] and Plutonium-239.[7] The operatic poem "Ideal Isotopes" is a canto of 239 lines in terza rima. The poem references two hundred and thirty nine religious deities associated with death, depicting the process of atomic fission as a kind of civil war taking place in a kind of meta-underworld, a mosaic of each human culture's respective planes of chthonic afterlife. In this chaotic realm, the totality of humanity's various supernatural figures, each one invented as a means of incarnating or mediating death, form alliances and wage war in order to lay claim to a portion of the influx of souls caused by the atomic age.

The poem "Critical Mass" unfolds a scenario in which a worker, preparing to leave "Fermi's pile"—a graphite reactor used to initiate and study nuclear chain reactions—experiences visions of various objects and structures, analogous to the pile.

The poem "Thuringa" describes Nazi efforts to create a tactical nuclear weapon towards the end of WWII. This poem is inspired by the article "Hitler 'tested small atomic bomb'" by Ray Furlong.[8]

The following long poem "Trinity"[9] forms its own section in this book in order to emphasize the significance of the events it describes. This poem chronicles the first successful test of a nuclear weapon, using imagery inspired by photographs of nuclear tests.

The next section, "Ghosts of Los Alamos,"[10] contains two poems. The first, "Valles Caldera," introduces the setting of the caldera—a collapsed, ancient super volcano located in New Mexico's Jemez Mountains—going on to describe the uranium enrichment process. The poem "The Demon Core" pays tribute to poet Michael Lista's book *Bloom*[11] and is dedicated to him and to artist Aaron Tucker. This poem recounts an incident where several scientists were exposed to astronomically high levels of radiation.

Moving from written to visual poems, the section "Industrial Complex,"[12] dedicated to poet Christopher Dewdney, responds to the visual collages that appear in his 1973 poetry collection *The Palaeozoic Geology of London, Ontario.*[13]

The poem "The Encore for Dr. Faustus" includes one of my original artworks, while "The Sleep of Reason" includes two original artworks drawn by Nicole Pucci for inclusion in this book. My most grateful thanks to her for this contribution to *The Manhattan Project.*

This section is also composed, in part, with the use of public domain images. "The Encore for Dr. Faustus" includes the 1946 photograph "Tickling the Dragon's Tail," showing a recreation of the Slotin incident.[14]

"The Dawn of Our Fall" includes an image of an electron,[15] a photo of the 1945 Trinity Site explosion,[16] and the etching *The Inferno, Canto 32* by Gustav Dore, completed sometime in the 1860s.

"The Sleep of Reason" Includes the etching *The sleep of reason produces monsters (No. 43)* by Francisco Goya, 1799, and a photo of the Trinity Nuclear Test, 1945,[17] while J. Robert Oppenheimer's eyes stare out from the bottom of the page.

"The Court of Fates" includes a photograph of Cerenkov radiation surrounding the underwater core of the Reed Research Reactor,[18] a crystal from *Rocks and Minerals: a Guide to Minerals, Gems, and Rocks, 400 Illustrations in Color,*[19] statues of Marie Curie[20] and The Virgin of Nagasaki,[21] and Jules Lefebvre's 1882 painting *La boite de Pandore.*

The next section, "Military Incidents,"[22] takes its titles from the United States military's nuclear incident terminology, a lexicon of code phrases used to track nuclear mishaps. Where an incident has not occurred, the section's poems instead respond to the military's metaphors with literalized scenarios.

The section "Rain of Ruin"[23] memorializes the victims of the Hiroshima and Nagasaki bombings.

The first poem in this section, "Clear Skies," erases the transcript from the Enola Gay, the military aircraft that dropped the two atomic bombs.

The second poem, "Testimony," unifies Japanese survivors' testimonies,[24] presenting a number of survivors' accounts as those of a single, unnamed speaker.

The third poem, "Operation Epsilon," quotes excerpts from the British government's transcription of the "Farm Hall Transcripts"—secret recordings made by British intelligence agents of German physicists who had worked on the Nazi nuclear program while they were being held under house arrest in 1946 at Farm Hall.

The following section, "Contamination,"[25] presents poetic accounts of incidents of nuclear pollution.

The poem "Christmas Island" re-presents Ted Blackwell's description of nuclear bomb tests that he witnessed.

"Plutonium Valley" dramatizes the United States' nuclear tests performed during Operation Plumbbob in the 1950s.

"The East Ural Reserve" comments on the Kyshtym disaster, where a chemical explosion released large quantities of radioactive material.

"The Argonne Incident," speculates upon the cause of an explosion that killed three nuclear plant workers in the United States.

"The Human Factor" explores the fallibility of both human-made instruments and reasoning, which conspired to cause the Three Mile Island partial meltdown.

"The Elephant's Foot" responds to the Russian K-19 and Chernobyl disasters, presenting a narrative involving a tourist on a trip to Pripyat.

"Rising Water" depicts elderly volunteers who aided in cleanup efforts following the Fukushima Daiichi meltdown.

The next section, "Doomsday Machines,"[26] depicts three types of plausible nuclear weapons.

The poem "Salted Bombs" depicts a post-apocalyptic scenario, caused by the use of nuclear weapons designed in order to maximize fallout.

The next two poems, "Antimatter" and "Strangelet," are sonnets in which each word contains only letters found within the title of its corresponding poem.

The final poem, "Destroyers of Worlds,"[27] lists all nuclear weapons tests performed thus far, as well as all countries' respective reserves of nuclear weapons. This poem is inspired by Moez Surani's *Operations*,[28] which lists the code names for all UN operations, and is dedicated to him.

NOTES

1 Joyelle McSweeney, *The Necropastoral: Poetry, Media, Occults* (Ann Arbor: University of Michigan Press, 2015).

2 McSweeney, *The Necropastoral*, 7.

3 The frontispiece to this book combines a portion of the etching *The Inferno, Canto 32* by Gustav Dore, c. 1860, and Stefanina Hill, "Vector – Atom Symbol," n.d., https://www.123rf.com/photo_38732808_stock-vector-atom-symbol.html.

4 The section begins with the illustration "Nuclear Fission Chain Reaction," Wikimedia Commons, June 2017, https://commons.wikimedia.org/wiki/File:Nuclear_fission_chain_reaction.svg.

5 This section begins with the illustration of a nuclear weapon "The B28 Type Thermonuclear Bomb," Nuclear Weapon Archive, https://nuclearweaponarchive.org/Library/Brown/B28bomb.gif.

6 This poem includes the electron diagram "Electron shell 092 uranium" by Greg Robson, Wikimedia Commons, August 2005, https://commons.wikimedia.org/wiki/File:Electron_shell_092_uranium.png.

7 This poem includes the electron diagram "Electron shell 094 plutonium," by Greg Robson, Wikimedia Commons, August 2005, https://commons.wikimedia.org/wiki/File:Electron_shell_094_plutonium.png.

8 BBC, 14 March 15, 2015, http://news.bbc.co.uk/2/hi/europe/4348497.stm.

9 This section begins with the illustration "Radiation warning symbol," Wikimedia Commons, September 2015, https://en.wikipedia.org/wiki/File:Radiation_warning_symbol.svg.

10 This section begins with a design by Melina Cusano inspired by a portion of the seal of Los Alamos County, and a portion of a map of New Mexico's Jemez Mountains, "Specific obsidian sources in the Jemez Mountain area," National Park Service, https://www.nps.gov/parkhistory/online_books/pecos/cris/images/fig9-3.jpg.

11 Michael Lista, *Bloom* (Toronto: House of Anansi Press, 2010).

12 This section begins with an illustration based on Frank B. Robinson, "Psychiana advertisement," 1946, from "Miss Atomic Bomb: the A-Bomb in Popular Culture—Comics, Cakes and the Will of God," JF Ptak Science Books, n.d., https://longstreet.typepad.com/thesciencebookstore/2011/03/miss-atomic-bomb-the-a-bomb-in-popular-culture-comics-cakes-and-god.html.

13 Christopher Dewdney, *The Palaeozoic Geology of London, Ontario* (Toronto: Coach House Books, 1973).

14 "Tickling the Dragon's Tail," 1946, Wikimedia Commons, https://en.wikipedia.org/wiki/File:Tickling_the_Dragons_Tail.jpg.

15 Greg Robson, "Electron shell 082 Lead," Wikimedia Commons, April 2006, https://commons.wikimedia.org/wiki/File:Electron_shell_082_Lead.svg.

16 Berlyn Brixner, "Trinity Test Fireball 16ms," 1945, Wikimedia Commons, https://en.wikipedia.org/wiki/File:Trinity_Test_Fireball_16ms.jpg.

17 "'Trinity' Explosion at Los Alamos, Alamogordo, New Mexico," 1945, from "Photos from the Trinity Nuclear Test, 1945," Xpda.com, n.d., http://xpda.com/junkmail/junk217/trinitytest.htm.

18 National Research Centre, "Cerenkov Effect," 2007, Wikimedia Commons, https://commons.wikimedia.org/wiki/File:Cerenkov_Effect.jpg

19 *Rocks and Minerals: a Guide to Minerals, Gems, and Rocks, 400 Illustrations in Color* (New York: Golden Press, 1957).

20 Nihil Novi, "Sklodowska-Curie statue," July 2010, Wikimedia Commons, https://commons.wikimedia.org/wiki/File:Sklodowska-Curie_statue,_Warsaw.JPG.

21 "The Virgin of Nagasaki" is s statue presented to the Cathedral of Urakami by Pope Francis in 1920 which survived the 9 August 1945 bombing. See https://presentationsistersne.ie/the-virgin-of-nagaski/.

22 This section begins with a portion of an illustration by SGM Timothy Lawn, "Elephant walking a B-52, ducking SAMs and surviving 'the Hilton,'" May 2016, Defence Visual Information Distribution Service, https://www.dvidshub.net/image/2620880/elephant-walking-b-52-ducking-sams-and-surviving-hilton.

23 This section begins with an illustration based on figure 3.17 in *The Effects of Nuclear Weapons*, edited by Samuel Glasstone, June 1957, https://archive.org/details/TheEffectsOfNuclearWeapons1957/mode/2up.

24 These testimonies were originally collected in the program "Hiroshima Witness," with English transcriptions and translations entitled "The Voice of Hibakusha" available at the Atomic Archive website http://www.atomicarchive.com/Docs/Hibakusha/index.shtml.

25 This section begins with an illustration derived from the photograph "Hazmat suit c. 1918," Wikimedia Commons, https://en.wikipedia.org/wiki/File:Hazmat_suit_c1918.jpg.

26 This section begins with an illustration derived from http://cui.unige.ch/isi/sscr/phys/img/bomb.jpg.

27 This section begins with the illustration "Palden Lhamo," from L. Austine Waddell, *Tibetan Buddhism*, 1895, Wikimedia Commons, https://commons.wikimedia.org/wiki/File:Palden_Lhamo.jpg. The illustration for the final section, "Fallout," is derived from https://www.abomb1.org/images/enw77b2.gif.

28 Moez Surani, *Operations: 1946-2006*, January 2017, https://moezsurani.com/Operations-1946-2006.

ACKNOWLEDGMENTS

The Manhattan Project is dedicated to the victims of nuclear weapons, accidents, and mishaps.

My deepest thanks go out to Alison Cobra, Kirsten Cordingly, Melina Cusano, Helen Hajnoczky, Brian Scrivener, Aritha van Herk, and everyone else at the University of Calgary Press for helping to make this book a reality.

Thanks to derek beaulieu, Christian Bök, Stephanie Bolster, Sina Queyras, Kate Sterns, Darren Wershler, and to all of the students enrolled in Concordia University's English 672 class for their support and critical input over the course of this book's inception and subsequent revision.

Last of all, my heart goes out to my infallible partner, Nicole Pucci.

Ken Hunt's writing has appeared in *Chromium Dioxide*, *Free Fall*, *Matrix*, and *Rampike* magazines, and in chapbooks from The Blasted Tree, No Press, and Penteract Press. His poetry has also been anthologized in *The Calgary Renaissance*, published in 2016 by Chaudiere Books, and in *Concrete and Constraint*, published in 2018 by Penteract Press. For three years, Ken served as managing editor of *NōD Magazine*, and for one year, as poetry editor of *filling Station*. He is the founder of *Spacecraft Press*, a small press publisher of experimental writing inspired by science and technology. The LUMA Foundation published his first book of poetry, *Space Administration*, in 2014. His second and third books of poetry, *The Lost Cosmonauts* and *The Odyssey*, were published by Book*hug Press in 2018 and 2019, respectively. Ken holds an MA in English from Concordia University and is a PhD candidate at the University of Western Ontario.

BRAVE & BRILLIANT SERIES

SERIES EDITOR:
Aritha van Herk, Professor, English, University of Calgary
ISSN 2371-7238 (Print) ISSN 2371-7246 (Online)

Brave & Brilliant encompasses fiction, poetry, and everything in between and beyond. Bold and lively, each with its own strong and unique voice, Brave & Brilliant books entertain and engage readers with fresh and energetic approaches to storytelling and verse, in print or through innovative digital publication.

No. 1 · *The Book of Sensations* | Sheri-D Wilson
No. 2 · *Throwing the Diamond Hitch* | Emily Ursuliak
No. 3 · *Fail Safe* | Nikki Sheppy
No. 4 · *Quarry* | Tanis Franco
No. 5 · *Visible Cities* | Kathleen Wall and Veronica Geminder
No. 6 · *The Comedian* | Clem Martini
No. 7 · *The High Line Scavenger Hunt* | Lucas Crawford
No. 8 · *Exhibit* | Paul Zits
No. 9 · *Pugg's Portmanteau* | D. M. Bryan
No. 10 · *Dendrite Balconies* | Sean Braune
No. 11 · *The Red Chesterfield* | Wayne Arthurson
No. 12 · *Air Salt* | Ian Kinney
No. 13 · *Legislating Love* | Play by Natalie Meisner, with Director's Notes by Jason Mehmel, and Essays by Kevin Allen, and Tereasa Maillie
No. 14 · *The Manhattan Project* | Ken Hunt